Henry Hilliard Earl

Fall River and Its Manufactories 1803-1890

With valuable statistical tables

Henry Hilliard Earl

Fall River and Its Manufactories 1803-1890
With valuable statistical tables

ISBN/EAN: 9783337237189

Printed in Europe, USA, Canada, Australia, Japan

Cover: Foto ©berggeist007 / pixelio.de

More available books at **www.hansebooks.com**

Fall River

and its

MANUFACTORIES.

1803—1890.

With Valuable Statistical Tables,

COMPILED FROM OFFICIAL SOURCES

BY HENRY H. EARL, A. M.

Twelfth Edition.   Carefully Revised.

FALL RIVER, MASS.

GEORGE E. BAMFORD

PUBLISHER,

1890

Almy & Milne, Printers, Fall River, Mass.

CONTENTS.

SUMMARY

OF THE

MANUFACTURING STATISTICS

OF

FALL RIVER, MASS.

January, 1890.

No. of Corporations, - - -	40
Capital Stock (Incorporated,) -	$20,643,000
No. of Mills, - - - - -	65
No. of Spindles, - - - -	2,128,228
No. of Looms, - - -	49,586
No. of Employees, - - - -	21,750
Pay Roll per Week, - - -	$145,405
Weekly Production—Pieces, -	221,000
Yards of Cloth per Annum, -	597,850,000
Bales of Cotton per Annum, -	244,850
No. of Water Wheels, - - -	12
No. of Steam Engines, - - -	108
Total Horse Power, (W. Wheels, 1,405 H. P.) -	47,435
Tons of Coal per Annum, - -	174,750
Gallons of Oil per Annum, - -	200,650
Pounds of Starch per Annum, -	2,245,000

FALL RIVER, MASS. :

Its Location, Water Power and Growth in Cotton Manufacture.

FALL RIVER, the largest Cotton Manufacturing center of America, is a city and port of entry of Bristol County, Massachusetts. The city is compactly built, and rises somewhat abruptly from the eastern shore of Mount Hope Bay, an arm of Narragansett Bay.

The remarkable water power and the spacious harbor of Fall River, are such in their conjunction as few cities on all the seaboard enjoy, and have been prime factors in the industrial history of the place. The water power is derived from a small stream (Fall River) —whence the name of the city,—which has its source in, or is in reality the outflow of a chain of ponds lying two miles east of the Bay, covering an area of 3,500 acres, and having a length of about eight miles, and an average breadth of three-quarters of a mile. They are mostly supplied by perennial springs, though receiving the outlets of several other sheets of water. The extent of country drained is comparatively small,— the whole water-shed being not over 20,000 acres, and the quantity of power therefore is to be attributed to the springs alluded to, and to the great and rapid fall of the river, which in less than half a mile is more than 127 feet. The flow of the river is 121 1-2 cubic feet per second, or 9,841,500,000 Imperial gallons in a year of three hundred and ten days of ten hours each.

The remarkable advantages of this river as a mill-stream have been increased by building a dam at the outlet of the ponds, which gives the water an additional fall of two feet, and its lower banks are largely built up with great manufacturing establishments, which, singly or in groups, rapidly succeed each other. The river for nearly its entire length runs upon a granite bed, and for much of the distance is confined between high banks, also of granite· Differing therefore from most other water-powers, this one allows the entire space between its banks to be occupied, and most of the water wheels connected with the older factories are placed directly in the bed of the river. While there is an almost uniform and constant supply of water, it is never subject to excess, and an injury in consequence of a freshet has never yet been known. As the river is thus perfectly controllable, the mills have been built directly across it, the wheels placed in its bed, and yet from an excess of water no damage is to be apprehended. In later years, the breast-wheels employed in the older mills have been supplanted by the modern appliances of turbine wheels and steam power.

The successful management of these factories on the stream (the oldest organized in 1813) was the foundation of the Cotton Manufacturing industry of Fall River.

In the course of years, with the increase of wealth and skill in manufacture, and the entrance upon the stage of action of younger men, new projects were formed, (1868-73) and since the older mills occupied all available space upon the river banks, new situations were sought out and appropriated, first on the margin of the ponds to the south and east of the city, and of which the stream is the outlet, and afterwards in the northerly and southerly sections of the city, on the banks of the Taunton river and Laurel lake.

The number of incorporated companies for the manufacture of cotton goods is now forty, owning sixty-five mills, with an incorporated capital cf $20,643,000, but a probable investment of $40,000,000, containing 2,128,228 spindles, and 49,586 looms.

The statistics of 1889 report the total number of mills in the United States as 962, containing 284,863 looms and 14,060,314 spindles, manufacturing 865,270,350 yards of print cloths per annum. Of these, New England has 521 mills, containing 231,405 looms and 11,500,364 spindles, manufacturing 702,891,685 yards of print cloths. Fall River has thus nearly one-sixth of all the spindles in the country, and about one-fifth of those in New England, and manufactures over THREE-FIFTHS of all the print cloths.

The following table will show the number of spindles in the mills of Fall River at the close of each year respectively:

1865,— 265,328.	1875,—1,269,048.	1883,—1,713,836.
1866,— 403,624.	1876,—1,274,265.	1884,—1,688,692.
1868,— 537,416.	1877,—1,284,701.	1885,—1,742,884.
1870,— 544,606.	1878,—1,322,169.	1886,—1,795.254.
1871,— 788,138.	1879,—1,364,191.	1887,—1,823.472.
1872,—1,094,702.	1880,—1,390,830.	1888,—1,893,944.
1873,—1,212,694.	1881,—1,429,412.	1889,—2,128,228.
1874,—1,258,508.	1882,—1,678,016.	

While the principal business of Fall River consists in the production of print cloths, its industrial activity is also engaged in the bleaching and dyeing of cotton goods; the printing of calicoes by the American Printing Co., one of the largest establishments in the country, and noted, especially, for its indigo prints; in the manufacture of cotton and other classes of machinery; of cotton thread, woolen goods, comforters, felt hats, boots, shoes, and other products of smaller factories, for which the city has unsurpassed facilities.

The harbor formed at the mouth of Taunton river, is safe, commodious, easy of access, and deep enough for ships of the largest class. Its steamers are of world-wide renown for size, beauty, strength and speed.

The city is underlaid with extensive beds of granite, easily quarried, and largely employed for building purposes. An inexhaustible supply of the purest water, containing only 1.80 grains of solid matter per gallon, is found in the lakes just back of the city, which is conveyed to the inhabitants through pipes laid by and under the control of a board of water commissioners, elected by the city government. A stand-pipe, reservoir and fire-hydrants, together with a thoroughly organized, paid fire department, and electric fire alarm, afford ample provision against loss by fire.

Good order is maintained by an efficient and well-officered police force of 90 patrolmen, and the record of crime is less than in any other city of its size in the country.

Educational interests have not been neglected. The annual appropriations by the city amount to $175,000 for the support of the High school, with English, Classical and Mixed courses, seven Grammar schools, a Training school and numerous Intermediate and Primary schools.

The cause of education has recently received a new impulse, by the munificent gift in 1887, by a mother in honor of her deceased son, of the " B. M. C. Durfee High School," a stately granite structure, occupying an entire square, with astronomical observatory, tower, clock and chimes, perfect in architectural design, complete. in every detail, and fully equipped with the most modern appliances of intellectual and physical training.

The city is provided with a Free Public Library of upwards of 36,000 volumes, and several Circulating Libraries, well supplied with the most recent publications, newspapers and magazines, and accessible to all. There are also numerous private and society libraries and local book clubs.

There are seven weekly and four daily papers.

Of churches there are twenty five Protestant and ten Roman Catholic, all well arranged and commodious, and with educated and talented preachers. Mission Schools shedding the kindly influence of Christianity here and there, have been established in various parts of the city, and under the care of devoted and self-sacrificing teachers, have continued from year to year with growing numbers and in creasing usefulness.

Fall River has sixty-seven lodges, distributed among Masons, Odd Fellows, Knights, Foresters, etc., seven Temperance Societies, and numerous benevolent organizations conveying aid and comfort to the needy and deserving, always to be found in every community.

. There are seven banks with an aggregate capital of $2,050,000, and a surplus account of $922,490, and four Savings Banks, carefully and faithfully conducted, having $12,132,658 deposits belonging to 26,595 depositors. A Loan and Trust Company and three Co-operative Banks aid materially in facilitating financial transactions.

The city is lighted with gas and electric lights ; its streets are generally wide and many of them well shaded; its private dwellings are neat and comfortable, and some of them of the finest architectural beauty. The winters are generally mild, being tempered by the Gulf Stream, while the summer is moderated by fresh breezes from the adjacent waters. The city is exceptionally healthy and entirely free from malaria. The trunk lines and many of the lateral branches of a complete system of sewers have been completed; a few of the principal streets have been paved; lines of horse-railroads radiate from the center to the suburbs, giving a convenient and ready means of transit. Public Parks have been purchased and laid out in different sections of the city and beautiful drives, giving extended views of the neighboring country, abound in the outskirts.

Fall River is often called the " Border City," because lying on the very borders of the State. Previous to 1862, a part of the city was in Massachusetts and a part in Rhode Island, the dividing line of the two States running through the southern center of the city. In that year, however, the boundary line was removed two miles south, and Fall River, Mass., and Fall River, R. I., became one, thus bringing the whole city into one municipality.

Fall River is 49 miles south of Boston, 183 miles north-east of New York, 17 miles south of Taunton, 18 miles south-east of Providence, 14 miles west of New Bedford, and 18 miles north of Newport. Daily lines of steamers connect Fall River, Providence, Newport and New York, while three lines of railways give ample passenger and freight communications inland.

Fall River was formerly a part of Freetown, and was incorporated as a separate town in 1803. Its name was soon after changed to Troy, but in 1834 its old appellation was restored. Its Indian name was "Quequetcant," signifying the "PLACE of falling water," and that of the river "Quequechan," which signifies "falling water" or "quick-running water," hence its appropriate name of Fall River; "Watuppa," the Indian name of the ponds on the east and by which they are still called, signifies "boats" or the "place of boats."

Fall River was incorporated a City in 1854.

Indian Names of Fall River and Vicinity.

ANNAWAN—1600 (?)-1676, "An officer," A Wampanoag, one of King Philip's most famous Captains.

CANONICUS—1557-(?)-1647, Chief of the Narragansetts; a friend of Roger Williams,

CONANICUT—Indian name of an Island in Narragansett Bay

CORBITANT—1590-(?)-1624. Sachem of Pocasset Tribe; chief residence at Gardner's Neck, Swansea.

KING PHILIP—1628-(?)-1676. English name of Metacomet, youngest son of Massasoit, and his successor, in 1662, as Chief of the Wampanoags.

MASSASOIT—1581-1661. Sachem of the Wampanoags and Chief of the Indian Confederacy formed of tribes in Eastern Massachusetts and Rhode Island. A staunch friend of the English.

METACOMET—Indian name of King Philip, second son of Massasoit.

MONTAUP—"The Head." Indian name of Mount Hope.

NARRAGANSETT—"At the Point." Indian tribe on the west side of Narragansett Bay.

NIANTIC—"At the River Point." Sub-tribe of the Narragansetts.

POCASSET—"At the opening of the Strait," i. e., Bristol Ferry into Mount Hope Bay. Indian name of the territory, now including Fall River and Tiverton.

QUEQUETEANT—"The PLACE of falling water." Indian name of Fall River.

QUEQUECHAN—" It leaps or bounds." Indian name of the stream—Fall River—signifying falling water or quick running water.

SAGAMORE—"A leader." Title of Indian Chief.

SEACONNET—" At the Sea Opening"—Indian Name of Little Compton.

TECUMSEH—1770-1813. Chief of the Shawnees : distinguished for his eloquence, bravery and manly virtues, Prominent on the Western frontier in the war of 1812.

WAMPANOAG—" East landers," i. e., east of Narragansett Bay. Indian tribe dwelling north and east of Narragansett Bay, west of Mount Hope Bay.

WAMSUTTA—1625-(?)-1662. English name, Alexander. Eldest son and successor of Massasoit in 1661.

WATUPPA—"Boats or the place of boats." Name of Ponds east of the city.

WEETAMOE—1620-(?)-1676. " Wise, shrewd, cunning." Daughter and successor of Corbitant as Sachem of the Pocasset tribe; residence at Fall River; drowned while crossing Slade's Ferry.

POPULATION—1810-1890.

POPULATION OF FALL RIVER AT VARIOUS TIMES

1810.... 1,296		1873.............38,464
1820.............1,594		1874.............43,289
1830............4,159		1875.............45,160
1840............6,738		1876 44,356
1845...........10,290		1877.............45,113
1850...........11,170		1878.............48,494
1855...........12,680		1879.............46,909
1860...........13,240		1880............ ...47,883
1862*...........17,461		1881.............49,049
1865...........17,525		1882.............50,487
1866...........19,262		1883.... 52,558
1867..........21,174		1884.............54,001
1868.... 23,023		1885.............56,863
1869.........25,099		1886.............59,026
1870...........27,191		1887..........63,961
1871...........28,291		1888....... 63,715
1872..............34,835		1889............ ...68,774

*The increase in population in 1862 was owing to the annexation of the Town of Fall River, R. I., which contained a population of about 3,590.

POPULATION BY WARDS AND PRECINCTS CORRECT-
ED TO MAY 1, 1889.

					Wards.				
Prec't.	1	2	3	4	5	6	7	8	9
A.	4,544	2,713	3,417	4,283	3,449	5,660	2,762	1,917	3,903
B.	6,404	4,363	3,516	3,921	3,620	4,907	2,527	1,167	5,701

OFFICIAL RETURN OF POLLS—MAY 1, 1889.

Prec't.	1	2	3	4	5	6	7	8	9
A.	1,076	739	906	959	975	1,336	698	513	985
B.	1,543	1,167	966	1,011	1,060	1,177	705	323	1,327

VALUATION, &c., 1854-1889.

Year.	Valuation.	Tax.	Amt. Raised by Taxation.	No. Polls.
1854	$8,939,215	$5.80	$56,523.70	3,117
1855	9,768,420	5.60	59,425.15	3,148
1860	11,522,650	7.40	90,124.61	3,238
1865	12,134,990	16.50	209,272.20	4,461
1870	23,612,214	15.30	374,753.22	6,743
1871	29,141,117	13.00	392,974.15	7,070
1872	37,841,294	12.00	471,835.53	8,870
1873	47,416,246	13.00	636,451.61	10,020
1874	49,995,110	12.80	662,486.11	11,119
1875	51,401,467	14.50	768,464.37	11,571
1876	48,920,485	15.20	764,629.41	10,519
1877	47,218,320	15 50	753,735.96	10,926
1878	42,329,730	17.50	739,518.48	11,564
1879	38,173,510	18.00	689,370.32	11,678
1880	39,171,264	18.00	702,088.91	12,008
1881	41,119,761	19.00	777,546.46	12,091
1882	43,421,970	18.80	813,490.93	12,881
1883	45,540,835	18.40	836,697.38	12,871
1884	45,798,860	18.80	859,013.53	13,212
1885	45,234,150	18.80	851,952.59	14,066
1886	45,111,705	18.80	852,014.82	14,677
1887	45,576,175	18.40	847,442.47	16,198
1888	46,504,585	17.40	862,823.98	16,319
1889	49,839,641	17.80	931,667.01	17,466

In 1840, the number of taxable polls was 1,603. The valuation of real estate was $1,678,603; of personal estate, $1,310,865; total, $2,989,468.

The Highest and Lowest Price in Print Cloths.

1850-1889.

The following table gives the Highest and Lowest price of Print Cloths from the year 1850 to 1889.

Year.	Highest.	Lowest.	Year.	Highest.	Lowest.
1850	5¾c	5c	1870	8¼	6½
1851	5¼	4¼	1871	8	6½
1852	5¾	4¼	1872	9	7⅜
1853	6⅜	6	1873	7½	5¾
1854	6¼	5½	1874	6⅛	5¼
1855	5½	4¾	1875	6¾	4⅜
1856	5¾	5	1876	4⅞	3⅝
1857	6⅛	5⅞	1877	5¼	3⅝
1858	6	5	1878	4	3⅛
1859	5⅞	5½	1879	4½	3 3-16
1860	5¾	4⅞	1880	5.87-100	3¾
1861	9	4¼	1881	4¼	3¾
1862	14½	7	1882	3.95-100	3⅝
1863	19	10¾	1883	3 13-16	3.44-100
1864	38½	16¼	1884	3.62-100	3.08-100
1865	27¼	10	1885	3¼	2.98-100
1866	19½	11¼	1886	3½	3⅛
1867	12	6⅝	1887	3⅝	3¼
1868	9⅝	6½	1888	4	3½
1869	9½c	7⅛c	1889	4 1-16	3½

EARLIEST COTTON MILLS IN U. S.

Bridgewater, Mass. (spinning mill)	1787
Beverly, Mass., do	1789
Slater, Pawtucket, R. I., do	1791
Bass River, Beverly, Mass., do	1801
New Ipswich, N. H., do	1804
Rehoboth, Mass., do	1805
New Ipswich (No. 2,) do	1807
Medway, Mass., do	1807
Fitchburg, Mass., do	} about
Waltham, Mass., do	} 1807-9
Brunswick, Me., do	1809
Waltham, first for both spinning & weaving,	1814

Spindles of United States in 1807—4,000.

WEEKLY PRODUCTION

OF PRINT CLOTH MILLS IN FALL RIVER.

Corporation.	No. of Mills.	Weekly Production, in Pieces.
American Linen Co.	2	10,500 Pieces.
Annawan Manufactory.	1	1,000
Barnard Manufacturing Co.	1	5,000
Border City Mfg. Co.	3	6,500
Chace Mills.	1	7,000
Cornell Mills.	1	6,000
Durfee Mills.	3	15,000
Fall River Iron Works Co.	1	7,500
Fall River Manufactory.	1	3,500
Flint Mills.	1	7,000
Granite Mills.	2	11,000
Hargraves Mills.	1	5,500
Laurel Lake Mills.	1	5.000
Mechanics' Mills.	1	7,500
Merchants' Manufacturing Co.	2	12,500
Metacomet Mfg. Co.	1	3,500
Narragansett Mills.	1	5,000
Osborn Mills.	2	6,500
Pocasset Manufacturing Co.	4	3,000
R. Borden Manufacturing Co.	2	12,000
Robeson Mills.	1	3,500
Sagamore Mfg. Co.	2	10,000
Seaconnet Mills,	1	6,000
Shove Mills.	2	9,000
Slade Mills.	1	5,500
Stafford Mills.	2	13,000
Tecumseh Mills.	2	6,000
Troy C. & W. Manufactory.	2	5,000
Union Cotton Mfg. Co.	3	8,000
Wampanoag Mills.	2	9,500
Weetamoe Mills.	1	5,000
		221,000

NOTE.—Most Print Cloth mills also manufacture "Convertibles," as cambrics, cheese cloths, inner linings, low grade bleached goods, etc.

FALL RIVER PRINT CLOTH MARKET.

For the Year Ending Aug. 31, 1888.

Week Ending.	Stock on Hand.	Sales of Week.	Price per Yard.	Price Md'lg Cott'n in N. York.
1887.	*Pieces.*			
Sept. 3	87,000	212,000	3¼	10
10	89,000	120,000	3¼	10⅛
17	72,000	142,000	3¼	9¾
24	64,000	137,000	3 5-16	9¾
Oct. 1	74,000	190,000	3 5-16	9½
8	74,000	179,000	3 5-16	9½
15	85,000	149,000	3 5-16	9½
22	39,000	272,000	3 5-16	9⅝
29	32,000	187,000	3¼	9⅝
Nov. 5	48,000	168,000	3¼	9⅝
12	39,000	261,000	3 5-16	10⅜
19	16,000	188,000	3⅜	10⅜
26	29,000	100,000	3 7-16	10½
Dec. 3	31,000	272,000	3 7-16	10⅝
10	31,000	163,000	3 7-16	10½
17	19,0 0	116,000	3 7-16	10⅝
24	3,000	282,000	3½	10⅝
31	2,000	192,000	3 9-16	10½
1888.				
Jan. 7	5,000	183,000	3⅝	10½
14	7,000	104,000	3¾	10½
21	8,000	114,000	3¾	10½
28	3,000	188,000	3⅞	10⅝
Feb. 4	5,000	228,000	4	10⅝
11	2,000	66,000	4	10⅝
18	8,000	57,000	4	10⅝
25	16,000	66,000	4	10½
Mch. 3	1,000	90,000	3¾	10½
10	1,000	68,000	3⅝	10⅛

FALL RIVER PRINT CLOTH MARKET.

For the Year Ending Aug. 31, 1888.

Week Ending.	Stock on Hand.	Sales of Week.	Price per Yard.	Price Md'lg Cott'n in N. York.
1888	Pieces.			
Mar. 17	.....	62,000	3⅝	10⅛
24	3,000	85,000	3⅝	10
31	......	98,000	3⅝	9⅞
April 7	28,000	113,000	3 9-16	9¾
14	10,000	263,000	3½	9¾
21	26,000	260,000	3½	9¾
28	35,000	219,500	3½	9¾
May 5	30,000	275,000	3½	10
12	22,000	264,000	3⅝	10
19	10,000	174,000	3 11-16	10
26	10,000	339,000	3¾	10
June 2	6,000	424,000	3¾	10
9	8,000	159,000	3¾	10
16	5,000	263,000	3⅞	10⅛
23	5,000	75,000	4	10¼
30	13,000	51,000	4	10¼
July 7	12,000	49,000	4	10¼
14	18,000	55,000	4	10½
21	20,000	60,000	3⅞	10⅝
28	1,000	110,000	3⅞	10¾
Aug. 4	......	163,000	3 13-16	10⅞
11	......	147,000	3⅞	10⅞
18	......	256,000	3¾	11⅜
25	......	301,000	3 13-16	10⅝

Cotton Crop 1887-88.........7,046,833 Bales.

FALL RIVER PRINT CLOTH MARKET.

For the Year Ending Aug. 31, 1889.

Week Ending.	Stock on Hand.	Sales of Week.	Price per Yard.	Price Md'lg Cott'n in N. York.
1888.	Pieces.			
Sept. 1	5,000	276,000	3⅞	11
7	2,000	312,000	3⅞	10 7-16
15	10,000	124,000	3⅞	10 7-16
22	7,000	256,000	3 13-16	10 7-16
29	14,000	248,000	3¾	10 7-16
Oct. 6	8,000	315,000	3¾	10 7-16
13	12,000	184,000	3¾	9 11-16
20	8,000	186,000	3¾	9¾
27	12,000	124,000	3 13-16	9 13-16
Nov. 3	17,000	43,000	3 13-16	9 13-16
10	22,000	202,000	3 13-16	10
17	24,000	151,000	3 13-16	10
24	7,000	382,000	3⅞	10
Dec. 1	6,000	95,000	3 15-16	9⅞
8	1,000	88,000	3 15-16	9⅞
15	10,000	117,000	3⅞	9⅞
22	......	149,000	3⅞	9¾
29	5,000	37,000	3 15-16	9¾
1889.				
Jan. 5	4,000	95,000	3 15-16	10 13-16
12	......	111,000	3 15-16	9⅞
19	......	380,000	4	9 15-16
26	......	166,000	4 1-16	9⅞
Feb. 2	......	297,000	4 1-16	9 15-16
9	......	107,000	4	10⅛
16	3,000	86,000	4	10
23	5,000	211,000	3 15-16	10⅛
Mar. 2	3,000	174,000	3 15-16	10 3-16
9	13,000	59,000	3 15-16	10 3-16

FALL RIVER PRINT CLOTH MARKET.

For the Year Ending Aug. 31, 1889.

Week Ending.	Stock on Hand.	Sales of Week.	Price per Yard.	Price M'dlg Cott,n in N. York.
1889.	Pieces.			
Mar. 16	7,000	84,000	4	10¼
23	......	202,000	4	10⅛
30	3,000	51,000	3 15–16	10 3–16
April 6	5,000	206,000	3⅞	10¼
13	22,000	320,000	3¾	10½
20	11,000	312,000	3¾	10 13–16
27	10,000	102,000	3¾	10 13–16
May 4	25,000	294,000	3¾	11 3–16
11	7,000	268,000	3⅞	11
18	8,000	184,000	3⅞	11 1–16
25	17,000	70,000	3⅞	11⅛
June 1	29,000	85,000	3⅞	11⅛
8	25,000	183,000	3 13–16	11 3–16
15	12,000	186,000	3⅞	11⅛
22	14,000	76,000	3⅞	11
29	8,000	203,000	3 15–16	11
July 6	8,000	43,000	4	11⅛
13	8,000	29,000	4	11¼
20	11,000	33,000	3 15-16	11¼
27	7,000	66,000	3 15–16	11 5–16
Aug. 3	7,000	66,000	3 15–16	11 5–16
10	7,000	238,000	3¾	11 5–16
17	10,000	242,000	3¾	11 5–16
24	10,000	94,000	3¾	11½
31	14,000	53,000	3¾	11½

Cotton Crop 1888-89,........6,938,290 Bales.

STATISTICS OF COTTON MANUFACTORIES

	Corporations.	Treasurer.
1	American Linen Co.,	Philip D. Borden,
2	Annawan Manufactory,	Thomas S. Borden,
3	Barnaby Mfg. Co.,	Stephen B. Ashley,
4	Barnard Manufg. Co.,	Nathaniel B. Borden,
5	Border City Mfg. Co.,	Edward L. Anthony,
6	Bourne Mills,	George A. Chace,
7	Chace Mills,	Joseph A. Baker,
8	Conanicut Mills,	Crawford E. Lindsey,
9	Cornell Mills,	John W. Hargraves,
10	Crescent Mills,	Benjamin M. Warren,
11	Davol Mills,	George H. Hills,
12	Durfee Mills.	David A. Brayton, Jr.,
13	F. R. Iron Works Co.,	Edward L Griffin,
14	Fall River Manufactory,	Holder B. Durfee,
15	Fall River Merino Co.,	Joseph Healy,
16	Flint Mills,	Wm. S. Potter,
17	Globe Yarn Mills,	Arnold B. Sanford,
18	Granite Mills,	Charles M. Shove,
19	Hargraves Mills,	Seth A. Borden,
20	King Philip Mills,	Simeon B. Chase.
21	Laurel Lake Mills,	Abbott E. Slade,
22	Mechanics Mills,	Horatio N. Durfee,
23	Merchants Manufg. Co.,	Andrew Borden,
24	Metacomet Manufg. Co.,	Thos. S. Borden,
25	Narragansett Mills,	James Waring,
26	Osborn Mills,	Joseph Healy,
27	Pocasset Manfg. Co.,	Bradford D. Davol,
28	Quequechan Mills,	Lessees,
29	Richard Borden Mfg. Co.,	Richard B. Borden,
30	Robeson Mills,	Clarence M. Hathaway,
31	Sagamore Mfg. Co.,	Hezekiah A. Brayton,
32	Seaconnet Mills,	Edward A. Chace,
33	Shove Mills,	Cyrus C. Rounseville,
34	Slade Mills,	Henry S. Fenner,
35	Stafford Mills,	Effingham M. Cock,
36	Tecumseh Mills,	Frank H. Dwelly,
37	Troy C. & W. Manuf'y,	Richard B. Borden,
38	Union Cotton Mfg. Co.,	Thos. E. Brayton,
39	Wampanoag Mills,	Walter C. Durfee,
40	Weetamoe Mills,	William Lindsey.

	Capital.	Spindl's	Looms. 32 & under	Over 32 in.	Style of Goods.
1	$800,000	85,568	2,043	.....	Print Cloths.
2	160,000	10,064	232	.....	" "
3	400,000	16,012	250	250	FineCol'dCot'nGds
4	330,000	33,760	816	102	P. C. & Specialties.
5	1,000,000	118,016	695	2,005	" & Wide Goods.
6	400,000	43,008	950	256	Odd Counts.
7	500,000	50,200	975	275	P.C.& Wide Goods.
8	120,000	16,172	.....	370	Wide Fine Goods.
9	400,000	42,000	300	660	P. C.& Odd Counts.
10	500,000	36,048	250	692	Fancy Woven Gds.
11	300,000	35,304	600	324	Sheet'gs & Fancies.
12	500,000	118,960	1,856	982	P. C.& Odd Counts.
13	500,000	44,800	1,200	.....	Print Cloths.
14	180,000	28,672	730	.....	" "
15	125,000	2,160	24	24	Merino Und'wear.
16	580,000	42,160	288	876	Goods to Order.
17	900,000	84,000	.....	....	Yarns, Fine & M. C'ts.
18	400,000	81,304	1,702	200	P. C.& Odd Counts.
19	400,000	37,632	328	588	Goods to Order.
20	1,000,000	103,440	48	2,252	Fine G'ds & Lawns
21	400,000	35,008	688	192	P. C. & Wide Goods
22	750,000	58,016	1,336	.....	Print Cloths.
23	800,000	89,888	2,055	100	P. C.& Odd Counts.
24	288,000	27,856	464	152	Goods to Order.
25	400,000	33,392	733	100	P.C.& Corset Jeans
26	600,000	70,200	1,017	831	P.C. Lwns&WdGds
27	800,000	68,084	694	881	Sh'gs, Twills,F'cies
28	....	14,104	.....	.....	Yarns—20s @36s.
29	800,000	84,904	1,100	960	P. C. W. Gds & F'cies.
30	260,000	24,096	600	.....	Print Cloths.
31	900,000	89,904	1,576	600	P.C.& Wide Goods.
32	400,000	35,280	926	.....	Print Cloths.
33	550,000	59,712	1,500	.....	P.C. & Odd Counts.
34	550,000	44,000	875	100	" & Wide Goods
35	800,000	82,496	1,780	324	Print Cloths.
36	500,000	46,704	949	229	P.C. & Wide Goods,
37	300,000	43,072	32	900	" & To Order.
38	750,000	91,152	1,072	1,131	" Wide Goods.
39	750,000	67,000	705	975	' & Staples.
40	550,000	34,080	860	.....	" Odd Counts.

$20,643,000 2,128,228 32,255+17,331=49,586 Looms.

STATISTICS OF COTTON MANUFACTORIES

	Corporations.	No. Mills.	Location.
1	American Linen Co.,	2	Ferry Street,
2	Annawan Manufactory,	1	Annawan Street,
3	Barnaby Mfg. Co.	1	Quequechan St.,
4	Barnard Mfg. Co.,	1	Quequechan St.,
5	Border City Mfg. Co.,	3	North Main St.,
6	Bourne Mills,	1	Laurel Lake,
7	Chace Mills,	1	Rodman Street,
8	Conanicut Mills,	1	Bay Street,
9	Cornell Mill.	1	Alden Street,
10	Crescent Mills,	1	Eight Rod Way,
11	Davol Mills,	2	Hartwell Street,
12	Durfee Mills,	3	Pleasant Street,
13	Fall River Iron Works Co.	1	Ferry Street.
14	Fall River Manufactory,	1	Pocasset Street,
15	Fall River Merino Co.,	1	Alden Street,
16	Flint Mills,	1	Alden Street,
17	Globe Yarn Mills,	3	Globe Street,
18	Granite Mills,	2	Twelfth Street,
19	Hargraves Mills.	1	Hargraves St.,
20	King Philip Mills,	3	Kilburn Street,
21	Laurel Lake Mills,	1	Broadway,
22	Mechanics Mills,	1	Davol Street,
23	Merchants Mfg. Co.,	2	Fourteenth St.,
24	Metacomet Manufg Co.,	1	Annawan Street,
25	Narragansett Mills,	1	North Main St.,
26	Osborn Mills,	2	Tower Street,
27	Pocasset Mfg. Co.,	4	Pocasset Street,
28	Quequechan Mills,	1	Pocasset Street,
29	Richard Borden Mfg. Co.,	2	Rodman Street,
30	Robeson Mills,	1	Hartwell Street,
31	Sagamore Mfg. Co.,	2	North Main St.,
32	Seaconnet Mills,	1	East Warren St.
33	Shove Mills,	2	Shove Street,
34	Slade Mills,	1	Globe Street,
35	Stafford Mills,	2	County Street,
36	Tecumseh Mills,	2	Hartwell Street,
37	Troy C. & W. Manuf'y,	2	Troy Street.
38	Union Cotton Mfg. Co.,	3	Pleasant Street,
39	Wampanoag Mills,	2	Quequechan St.,
40	Weetamoe Mills,	1	Davol Street,

IN FALL RIVER, MASS.

	In-cor-p'd.	Bls. Cotton used per ann.	Yards of Cloth manufactured per annum.	No. Hands Empl'd.	Weekly Pay Roll.
1	1852	10,000	24,750,000	1,000	$5,400
2	1825	1,200	2,600,000	135	625
3	1882	1,200	....	400	3,000
4	1874	4,000	11,500,000	325	2,350
5	1880	12,000	27,500,000	1,200	7,500
6	1881	4,600	14,000,000	500	3,200
7	1871	6,200	15,500,000	475	3,300
8	1880	1,300	2,500,000	175	1,300
9	1889	5,000	14,000,000	425	3,100
10	1871	4,000	9,500,000	400	2,700
11	1867	4,000	7,000,000	425	2,600
12	1866	13,500	35,500,000	1,050	7,500
13	1825	6,500	17,000,000	575	3,300
14	1813	3,500	7,500,000	300	1,800
15	1875	800	1,750,000	150	800
16	1872	6,500	13,750,000	500	3,350
17	1881	12,000	....	750	5,000
18	1863	9,000	25,000,000	780	5,300
19	1888	5,000	14,500,000	400	2,600
20	1871	5,500	15,000,000	1,000	7,800
21	1881	4,500	13,000,000	365	2,400
22	1868	6,500	17,000,000	570	3,850
23	1867	10,000	27,000,000	825	5,600
24	1880	3,000	8,000,000	325	1,600
25	1871	4,000	11,000,000	325	2,200
26	1871	8,000	19,500,000	800	5,800
27	1822	9,500	23,000,000	750	5,000
28	....	1,500	....	75	480
29	1871	9,500	28,000,000	775	5,500
30	1866	3,000	7,500,000	275	1,700
31	1879	10,250	27,000,000	900	5,800
32	1884	5,000	14,000,000	400	2,500
33	1872	7,000	19,000,000	625	4,200
34	1871	5,000	13,500,000	400	2,600
35	1871	10,000	29,000,000	775	5,500
36	1866	5,200	13,500,000	400	3,000
37	1814	4,500	12,000,000	375	2,500
38	1879	10,500	25,000,000	900	5,900
39	1871	8,000	20,000,000	650	4,300
40	1871	4,100	12,000,000	375	2,450
		244,850	597,850,000	21,750	$145,405

STATISTICS OF COTTON MANUFACTORIES.

	Corporations.	Superintendents.
1	American Linen Co.,	John A. Collins,
2	Annawan Manufactory,	Benj. T. Almy,
3	Barnaby Mfg. Co.,	Geo. Cunningham,
4	Barnard Manuf'g Co.,	Wm. Hathaway,
5	Border City Mfg. Co.,	Gilbert P. Cuttle,
6	Bourne Mills,	Raymond Murray,
7	Chace Mills,	P. A. Mathewson,
8	Conanicut Mills,	Wm. H. Swift,
9	Cornell Mills,	Isaac H. Russell,
10	Crescent Mills,	Harry G. Baker,
11	Davol Mills,	Timothy Sullivan,
12	Durfee Mills,	{ E. P Emery, No. 1, 3, (Chas. C. Diman, No. 2
13	F. R. Iron Works Co.,	Sam'l E. Hathaway,
14	Fall River Manufactory,	Alfred H. Hartley,
15	Fall River Merino Co.,	Matt. C. Yarwood,
16	Flint Mills,	Joseph Shaw,
17	Globe Yarn Mills,	John A. Sanford,
18	Granite Mills,	James E. McCreery,
19	Hargraves Mills,-	John P. Bodge,
20	King Philip Mills,	Albert A. Sweet,
21	Laurel Lake Mills,	Horace W. Tinkham,
22	Mechanics Mills,	James C. Davol,
23	Merchants Manuf'g Co.,	John Gregson,
24	Metacomet Manuf'g Co.,	Benj. T. Almy,
25	Narragansett Mills,	John Harrison,
26	Osborn Mills,	Joseph Watters,
27	Pocasset Manuf'g Co.,	W. S. Whitney, S. & Agt.
28	Quequechan Mills,	—— ——
29	R. Borden Manuf'g Co.,	Herbert H. Shumway,
30	Robeson Mills,	Ellery B. Healy,
31	Sagamore Manuf'g Co.,	Geo. Whittaker,
32	Seaconnet Mills,	J.E. Cunneen, S.& Agt.
33	Shove Mills,	Benj. B. Kirk,
34	Slade Mills.	Daniel J Harrington,
35	Stafford Mills,	Samuel W. Hathaway,
36	Tecumseh Mills,	Robinson Walmsley,
37	Troy C. & W. Manuf'y,	Wm. E. Sharples,
38	Union Cotton Mfg. Co,	Edward Lynch,
39	Wampanoag Mills,	Roland R. Kelley,
40	Weetamoe Mills.	Richard Thackeray,

Engines No.	Horse Power	PER ANNUM. Tons Coal	Galls. Oil	Lbs. Starch.
1	5 1,500	7,500	10,000	100,000
2 {	2 200			
{	1 w.wh.140	1,000	1,000	12,000
3	2 530	3,000	1,700	......
4	2 750	3,000	3,200	40,000
5	4&1a 2,350	8,500	10,000	120,000
6	2 950	4,500	5,000	52,000
7	3 1,000	4,500	5,000	66,000
8	1 300	1,250	1.400	12,000
9	1 a 1,100	3,500	3,800	33,000
10	2 700	3,000	2,500	50,000
11	2 650	3,700	7,000	48,000
12	6 2,650	11,000	12,000	100,000
13	2 b 1,400	3,300	4,200	42,000
14 {	1 500			
{	2 w.wh.260	2,000	3,000	34,000
15	1 125	1,000	1,000	......
16	2 1,050	4,500	4,500	57,000
17	3&2a 2,000	6,200	4,300	......
18	4 1,500	7,000	8,000	100,000
19	1 a 1,000	3,300	3,500	33,000
20	4 1,600	7,000	8,000	55,000
21	3 800	3,000	3,000	40,000
22	2 a 1,350	4,200	4,500	75,000
23	5 1,700	7,500	7,500	110,000
24 {	1 500			
{	1 w.wh.150	2,200	3,000	35,000
25	1 a 1,000	3,000	3,500	50,000
26	2&1a 1,700	6,500	7,500	87,000
27 {	4 1,425			
{	5 w.wh.625	4,500	6,000	80,000
28 {	2 700			
{	1 w.wh. 80	1,000	1,000	......
29	2&1b 1,800	7,000	7,500	100,000
30	2 450	2,000	2,500	32,000
31	2&1a 2,400	7,000	7,500	100,000
32	2 650	3,500	3,500	35,000
33	4 1,200	5,000	5,500	65,000
34	3 800	4,000	3,500	60,000
35	2 a 1,850	5,000	6,500	100,000
36	2&1a 1,100	4,000	5,000	55,000
37 {	1 a 700			
{	2 w.wh.150	2,000	3,300	40,000
38	6 1,700	7,000	7,500	110,000
39	4 1,700	5.600	10,250	75,000
40	1 a 650	2.000	3,000	42,000
	47,435	174,750	200,650	2,245,000

W.Wh-Water Wheel, "a"-Compound Engine. "b"-Triple Expansion Engine.

PAY DAYS of the SEVERAL CORPORATIONS.

American Linen Co.,	Fridays.
American Printing Co.,	Saturdays.
Annawan Manufactory.	Saturdays.
Barnaby Manufacturing Co.,	Saturdays.
Barnard Manufacturing Co.,	Thursdays.
Border City Manufacturing Co.,	Thursdays·
Bourne Mills,	Fridays.
Chace Mills,	Fridays.
Conanicut Mills,	Fridays.
Cornell Mills,	Tuesdays.
Crescent Mills,	Fridays.
Davol Mills.	Fridays.
Durfee Mills,	Wednesdays
Fall River Bleachery,	Wednesdays.
Fall River Iron Works Co.,	Saturdays.
Fall River Manufactory,	Wednesdays.
Fall River Merino Co.,	Saturdays.
Flint Mills,	Tuesdays.
Globe Yarn Mills.	Fridays.
Granite Mills.	Fridays.
Hargraves Mills.	Fridays.
King Philip Mills,	Saturdays.
Laurel Lake Mills,	Saturdays.
Mechanics Mills,	Fridays.
Merchants Manfg. Co.,	Saturdays.
Metacomet Manfg. Co.,	Saturdays.
Narragansett Mills,	Fridays.
Osborn Mills,	Fridays.
Pocasset Manfg. Co.,	Fridays.
Quequechan Mills,	Thursdays.
R. Borden Manfg. Co.,	Saturdays.
Robeson Mills,	Thursdays.
Sagamore Manfg. Co.,	Thursdays.
Seaconnet Mills,	Saturdays.
Shove Mills,	Thursdays.
Slade Mills,	Saturdays.
Stafford Mills,	Fridays.
Tecumseh Mills,	Fridays.
Troy Cotton & Woolen Manfy.,	Saturdays.
Union Cotton Manfg. Co.,	Thursdays.
Wampanoag Mills,	Fridays.
Weetamoe Mills,	Fridays.

MISCELLANEOUS CORPORATIONS

In Fall River, Mass.

Corporations.	Incorporated.	Capital.
American Printing Co.,	1880	$750,000
Beattie Battery Z. & Elec. Co.,	1887	1,000,000
Bay State Shoe Fastening Co.,	1887	200,000
Border City Hotel Co.,	1888	150,000
Carr Metal Co.,	1889	50,000
Chase's Patent Elevator Co.,	1887	110,000
Crystal Spring Bleach. & D. Co.	1881	200,000
Edison Electric Illuminating Co.	1883	90,000
F. R. Bleachery,	1874	400,000
" Boot and Shoe Mfg. Co.,	1887	15,000
" Electric Light Co.,	1883	90,000
" Gas Works Co.,	1880	288,000
" Machine Co..	1880	96,000
F. R. Mill Supply Co.,	1889	10,000
" Mfrs. Mut. Insurance Co.,	1870	......
" & Prov. Steamboat Co.,	1880	96,000
" Railroad Co..	1874	200,000
" Spool & Bobbin Co.,	1878	21,000
" Warren & Prov. R. R. Co.	1857	150,000
Globe Street Railway Co.,	1880	300,000
Granite City Soap Co.,	1888	650,000
Jesse Eddy Man'f'g Co.,	1886	60,000
Kerr Thread Co.,	1888	1,000,000
Kilburn, Lincoln & Co..	1868	80,000
Manufacturers Gas Light Co.,	1880	50,000
Massasoit Manufacturing Co.,	1882	50,000
Old Colony R. R. Co.,	1844	12,166,800
" " Steamboat Co.,	1874	1,200,000
Union Belt Co.,	1871	48,000

Pay Day of the Several Corporations in Fall River.

The system of WEEKLY PAYMENTS was adopted by the Corporations of Fall River in 1880-81.

Cotton Industry of the United States.

Census of 1880.

Repoited by Edw. Atkinson, Boston, Special Agent.

States.	No. of Mills.	Capital.	Spindles.	Looms.
Alabama,	15	$1,186,500	49,432	863
Arkansas,	2	75,000	2,015	28
Connecticut,	81	20,100,500	933,540	18,161
Delaware,	9	879,571	48,858	786
Florida,	1	11,000	816	
Georgia,	41	6,363,657	199 578	4,390
Illinois,	2	240,000	4,860	24
Indiana,	4	1,090,000	33,396	776
Kentucky,	3	360,000	9,022	73
Louisiana,	2	195,000	6,096	120
Maine,	24	15,092,080	696,564	15,978
Maryland,	20	4,003,816	125,706	2,425
Mass.,	175	72,896,448	4,276,723	95,671
Michigan,	1	20,000	5,100	131
Mississippi,	6	951,140	18,568	644
Minnesota,	1	5,000	1,708	24
Missouri,	1	690,000	19,312	431
N. Hampshire	36	19,517,085	1,008,509	25,503
N. Jersey,	18	3,208,500	232,221	3.180
New York,	36	11,179,318	573,390	12,575
N. Carolina,	49	2,858,800	100,209	1,770
Ohio,	4	670,000	14,328	42
Pennsylvania	55	10,249,986	425,247	8,211
Rhode Island,	115	29,048,671	1,648,917	29,881
So. Carolina,	15	2,768,500	92,424	1,676
Tennessee,	16	1,140,600	39,236	806
Texas,	2	50,000	2,648	71
Utah,	1	20,000	432	14
Vermont,	7	936,096	55,081	1,180
Virginia.	7	1,115,100	44,340	1,322
Wisconsin,	2	202,500	10,240	400
	751	207,781,868	10,678,516	227,156

NOTE.—The above does not include Hosiery Mills, or any of the mills known as Woolen Mills, where Cotton may be a component material used in the manufacture.

SUMMARY OF SPINDLES OF THE WORLD.

SEPT. 1889.

	No. of Spindles.	Bales Cotton used Per Annum.
United States,	14,060,000	2,767,000
Great Britain,	43,500,000	3,770,000
Europe, (Cont.)	24,000,000	4,069,000
India,	2,763,000	889,000
	84,323,000	11,495,000

COTTON SPINNING IN UNITED STATES.

SEPT. 1889.

	Mills.	Spindles.	Looms.
United States,	962	14,060,314	284,863
Northern States,	709	12,755,967	260,057
Southern States,	252	1,304,347	24,806
New England,	521	11,500,364	231,405
Fall River,	65	2,128,228	49,586

	U. S.	N. E.	F. R.
Print Cloths M'f'd,	865,270,350	702,891,685	597,850,000

COTTON CENTERS OF THE N. E. STATES.

	Capital.	Spindles.	Looms.	Employees.
Lewiston Me.,	$5,550,000	318,734	7,374	6,082
Manchester,N.H.,	8,600,000	468,600	15,657	11,490
Lawrence, Mass.,	8,525,000	415,448	12,445	12,720
New Bedford, "	8,250,000	694,416	12,651	5,219
Lowell, "	18,600,000	1,004,822	29,975	27,140
Fall River, "	20,643,000	2,128,228	49,586	21,750
Providence and Blackstone Val'y	17,067,000	1,484,960	32,957	17,715

PRINT WORKS IN THE UNITED STATES.

	Name of Print Works.	Location.
1	Allen,	Providence, R. I.,
2	American,	Fall River, Mass.,
3	Arnold,	North Adams, Mass.,
4	Cocheco,	Dover, N. H.,
5	Dunnell,	Pawtucket, R. I.,
6	Eddystone,	Chester, Penn.,
7	Franklin,	Patterson, N. J.,
8	Freeman,	North Adams, Mass.,
9	Garner & Co.,	Haverstraw, N. Y.,
10	" "	Wappinger Falls, N. Y.,
11	Gloucester,	Gloucester, N. J.,
12	Hamilton,	Lowell, Mass.,
13	Hartel,	Holmesburg Jnct'n, Pa.
14	Manchester,	Manchester, N. H.,
15	Merrimack,	Lowell, Mass.,
16	Mystic,	Medford, Mass.
17	Pacific,	Lawrence, Mass.,
18	Passaic,	Passaic, N. J.,
19	Southbridge,	Southbridge, Mass.
20	Washington,	River Point, R. I.,

PRINT WORKS IN THE UNITED STATES.

	Capital of Incorporated Companies.	No. of Print'g Machines.	No. Pieces of Calico Pri't'd per Week.	No. Pieces of Print Cloths made per week by the Com'y.
1	$300,000	13	22,000	6,300
2	750,000	19	33,000	None.
3	150,000	10	17,000	6,500
4	1,500,000	13	23,000	15,000
5	700,000	11	19,000	None.
6	1,000 000	21	36,000	1,500
7	170,000	7	6,000	None.
8	150,000	12	20,000	2,000
9	Not Incor.	20	{ 50,000	{ 40,000
10	"	22		
11	144,000	12	18,000	None.
12	1,800,000	10	16,000	10,000
13	Not Incor.	5	7,000	None.
14	2,000,000	16	18,000	6,000
15	2,500,000	20	36,000	22,000
16	Not Incor.	4	6,000	None.
17	2,500,000	30	52,000	5,000
18	150,000	10	16,000	None.
19	12,000	5	Not Print'g	None.
20	Not Incor.	9	15,000	None.
		269	410,000	114,300

DIVIDENDS OF FALL RIVER MILLS.

CORPORATION.	Capital	Par value	Dividends, 1888-1889.	
American Linen Co.	$800,000	100	$90	15¼
Annawan Mfg. Co.	160,000	500		
Barnaby Mfg. Co.	400,000	100	8	11
Barnard Mfg. Co.	330,000	100	7½	8½
Bourne Mills	400,000	100	16	16
Border City Mfg. Co.	1,000,000	100	11	9
Chace Mills	500,000	100	7	8
Conanicut Mills	120,000	100	6	6
Crescent Mills	500,000	100	4	2
Davol Mills	300,000	100	2	8
Durfee Mills, (Private.)	500,000	100		
F. R. Iron W'ks Co. "	500.000	100		
Fall River Manufac'ry	180.000	100	12	12
Flint Mills	580,000	100	10	14
Globe Yarn Mills	900,000	100	8	8
Granite Mills	400,000	100	22	24
King Philip Mills	1,000,000	100	6	6
Laurel Lake Mills	400,000	100	8½	12
Mechanics Mills	750,000	100	6¼	7½
Merchants Mfg. Co.	800,000	100	7½	10
Metacomet Mfg. Co.	288,000	100	3	5
Narragansett Mills	400,000	100	8	8
Osborn Mills	600,000	100	6	6
Pocassett Mfg. Co.	800,000	100	8	8
Rich. Borden Mfg. Co.	800,000	100	6	8
Robeson Mills	260,000	100	6	7¼
Sagamore Mfg. Co.	900,000	100	13	10½
Seaconnet Mills	400,000	100	15	17
Shove Mills	550,000	100	6¼	8
Slade Mills	550,000	100	4¼	6
Stafford Mills	800,000	100	9	6
Tecumseh Mills	500.000	100	8	10
Troy C. & W. Mf'y	300,000	500	100	$120
Union Cotton Mfg. Co.	750,000	100	30	20
Wampanoag Mills	750,000	100	11½	17
Weetamoe Mills	550,000	100	3¼	6

Average rate for 1889 - - - 9.98 per cent.
Average rate for past 4 years - 8.66 "

ORGANIZATION

—OF—

CORPORATIONS.

LIST OF OFFICERS

—AND—

DATES OF ANNUAL MEETINGS.

JANUARY, 1890.

ORGANIZATION

OF

CORPORATIONS.

January, 1890.

American Linen Company.

PRESIDENT: John S. Brayton.
CLERK: Daniel E. Chace.
TREASURER: Philip D. Borden.
DIRECTORS: John S. Brayton, Richard B. Borden, Samuel M. Brown, Clark Shove, J. P. Prentiss, Jefferson Borden, Fall River; Horace M. Barns, Bristol.
Annual Meeting—1st Wednesday in November.

American Printing Company.

PRESIDENT: Matthew C. D. Borden.
CLERK AND TREASURER: Alphonso S. Covel.
DIRECTORS: M C.D.Borden, Cornelius N.Bliss, John Van Wormer, S. P. Marshall, New York; Alphonso S. Covel, Boston.
Annual Meeting—1st Wednesday in August.

Annawan Manufactory.

PRESIDENT: John S. Brayton.
CLERK AND TREASURER: Thomas S. Borden.
DIRECTORS: John S. Brayton, Richard B. Borden, Philip D. Borden, Thos. S. Borden, Holder B. Durfee.
Annual Meeting—1st Tuesday in August.

Barnaby Manufacturing Co.

PRESIDENT: Simeon B. Chase.
CLERK AND TREASURER: Stephen B. Ashley.
DIRECTORS: Simeon B. Chase, George H. Hawes, George H. Hills, Stephen B. Ashley, A. J. Chace, B. D. Davol, J. C.Borden; Wm. F. Draper, Hopedale ; Charles E. Barney, New Bedford.
Annual Meeting—Last Monday in April.

Barnard Manufacturing Co.

PRESIDENT: James M. Aldrich.
CLERK AND TREASURER: Nathaniel B. Borden.
DIRECTORS: James M. Aldrich, Simeon Borden, N. B. Borden, Robert T. Davis, L. Lincoln, S. B. Chase, B. D. Davol; Arnold B. Chace, Valley Falls; Wm. Huston, Providence; W. H. Gifford, No. Westport.
Annual Meeting—4th Thursday in October.

Beattie Battery Zinc & Electrical Co.

PRESIDENT : Benjamin Cook.
CLERK : Clarence Hale, Portland.
TREASURER : John Beattie, Jr.
DIRECTORS : Benjamin Cook, John Beattie, Jr.; Howes Norris, Cottage City; Arthur Lord, Plymouth.
Annual Meeting—1st Wednesday in May.

Bay State Shoe Fastening Co.

PRESIDENT : Weaver Osborn.
CLERK : Geo. B. French, Nashua.
TREASURER : James C. Brady.
DIRECTORS: Weaver Osborn, Geo. W. Slade; James C. Brady, Boston; Wendell H. Cobb, New Bedford ; Edmund Woodman, Portland ; Geo. H. Stearns, Manchester.
Annual Meeting—4th Thursday in January.

Border City Hotel Co.

PRESIDENT: George F. Mellen.
CLERK: Edwin H. Davis.
TREASURER: John W. Hargraves.
DIRECTORS: Geo. F. Mellen, Edwin H. Davis, John W. Hargraves.
Annual Meeting—4th Wednesday in August.

Border City Manufacturing Company.

PRESIDENT: John S. Brayton.
CLERK: Henry K. Braley.
TREASURER: Edward L. Anthony.
DIRECTORS: John S. Brayton, Chas. J. Holmes.
Thomas E. Brayton, Fall River ; J. A. Beauvais,
Chas. E. Barney, Gilbert Allen, New Bedford;
Francis A. Foster, A. S. Covel, Boston ; Geo. M.
Woodward, Taunton.
Annual Meeting—4th Wednesday in May.

Bourne Mills.

PRESIDENT : Jonathan Bourne, Jr.
CLERK AND TREASURER : George A. Chace.
DIRECTORS : Jonathan Bourne, Jr., Portland,
Ore.; George A. Chace, Lloyd S. Earle, Charles
M. Shove, Frank S. Stevens; Stephen A. Jenks,
Pawtucket; Nath. B. Horton, Rehoboth.
Annual Meeting—In October.

Carr Metal Co.

PRESIDENT: Robert L. Carr.
CLERK: Henry N. Carragher.
TREASURER: Parker Borden.
DIRECTORS: Robt. L. Carr, Parker Borden,
Henry N. Carragher, Matt. C. Yarwood, Geo. B.
Durfee, Robt. Nicholson, Wm. A. Dolan; Samuel
Seabury, Tiverton.
Annual meeting—1st Thursday in June.

Chace Mills.

PRESIDENT: Edward E. Hathaway.
CLERK AND TREASURER: Joseph A. Baker.
DIRECTORS: Edward E. Hathaway, George W.
Grinnell, Joseph A. Baker, Jerome C. Borden,
Adoniram J. Chace; Wm. A. Abbe, N. B.;
David P. Davis, Somerset.
Annual Meeting—In October.

Chase's Patent Elevator Co.

PRESIDENT: Clark Chase.
CLERK: Wm. M. Hawes.
TREASURER: Thomas Kieran.
DIRECTORS: Clark Chase, Frank S. Stevens,
Robert T. Davis, Holder B. Durfee, Simeon B.
Chase, Wm. M. Hawes, Wm. F. Hooper.
Annual Meeting—3d Wednesday in October.

Conanicut Mills.

PRESIDENT: Edmund W. Converse.
CLERK AND TREASURER: C. E. Lindsey.
DIRECTORS: E. W. Converse, Boston ; James
H. Chace, Providence; Wm. Lindsey, Crawford
E. Lindsey, Clarence A. Brown, Fall River; Chas.
E. Barney, New Bedford.
Annual Meeting—4th Wednesday in October.

Cornell Mills.

PRESIDENT: John D. Flint.
CLERK· James F. Jackson.
TREASURER: John W. Hargraves.
DIRECTORS: John D. Flint, Reuben Hargraves,
Thomas Hargraves, Daniel H. Cornell, Clark
Chase, Jas. F. Jackson; Stephen A. Jenks,
Pawtucket; Wm. F. Draper, Hopedale; Arthur L.
Kelley, Valley Falls; Myron A. Fish, Prov.;
Rodman P. Snelling, Boston; Cyrus Washburn,
Wellesley.
Annual Meeting—2d Tuesday in November.

Crescent Mills.

PRESIDENT: Benjamin Covel.
CLERK AND TREASURER: Benjamin M. Warren.
DIRECTORS: Benjamin Covel, N. Arnzen;
Alphonso S. Covel, Boston; Chas. E. Barney, N.
B.; Wm. H. Bent, Taunton.
Annual Meeting—2d Wednesday in November.

Crystal Spring Bleaching & Dyeing Co.

PRESIDENT : Robert Henry.
CLERK AND TREASURER : John P. Henry.
DIRECTORS : Robert Henry, Simeon B.Chase,
John P. Henry, Milton Reed, James W. Henry,
Leontine Lincoln.
Annual Meeting—In February.

Davol Mills.

PRESIDENT: Arnold B. Sanford.
CLERK AND TREASURER: George H. Hills.
DIRECTORS: A. B. Sanford, Jonathan Slade,
Simeon B. Chase, Geo. S. Eddy; Chas. R. Batt,
J. O. Wetherbee, Boston; W. S. Granger, Prov;
C. W. Haskins, John J. Hicks, New Bedford.
Annual Meeting—1st Monday in May.

Durfee Mills.

PRESIDENT: John S. Brayton.
CLERK AND TREASURER: David A. Brayton, Jr.
DIRECTORS: John S. Brayton, Hezekiah A. Brayton, David A. Brayton, Jr.; Bradford W. Hitchcock, New York.
Annual Meeting—4th Tuesday in October.

Edison Electric Illuminating Co.

PRESIDENT : J. C. Borden.
CLERK: Henry K. Braley.
TREASURER: Albert F. Dow.
DIRECTORS : Jerome C. Borden, Frank S. Stevens, Jas. B. Harley, W. S. Whitney, James P. Hillard, A. F. Dow; F. S. Hastings, New York.
Annual Meeting—3d Wednesday in October.

Fall River Bleachery.

PRESIDENT: George W. Dean.
CLERK: Geo. O. Lathrop.
TREASURER: Spencer Borden.
DIRECTORS: George W. Dean, Spencer Borden, Richard B. Borden, Bradford D. Davol: Thomas Bennett, Jr., Joseph A. Beauvais, New Bedford; John Waterman, Warren.
Annual Meeting—3d Thursday in May.

Fall River Boot & Shoe Manfg. Co.

PRESIDENT: Wm. A. Dolan.
CLERK: John E. Sullivan.
TREASURER: Francis Quinn.
DIRECTORS: Wm. A. Dolan, Daniel D. Sullivan, Quinlan Leary, Daniel Murphy, Francis Quinn.
Annual Meeting—2d Tuesday in January.

Fall River Electric Light Co.

PRESIDENT: Marsden J. Perry.
CLERK: Fred O. Dodge.
TREASURER: Owen Durfee.
DIRECTORS: M. J. Perry, Prov; Frank S. Stevens, John D. Flint, W. H. Hathaway, F. O. Dodge, Edw. L. Anthony; Wm. B. Hosmer, Boston.
Annual Meeting—2d Monday in March.

Fall River Gas Works Company.

PRESIDENT : John S. Brayton.
CLERK AND MANAGER : Geo. P. Brown.
TREASURER: Edward C. Lee.
DIRECTORS : John S. Brayton, Elihu Andrews,
Geo. P. Brown; Sam'l T. Bodine, Randall
Morgan; Henry Levis, Phila.; David Paton, N. Y.
Annual Meeting—Last Wednesday in July.

Fall River Iron Works Co.

PRESIDENT: Matthew C. D. Borden.
CLERK AND TREASURER: Edward L. Griffin.
DIRECTORS: M.C.D. Borden, Cornelius N.Bliss,
S. P. Marshall, New York; A. S. Covel, Boston.
Annual Meeting—1st Tuesday in August.

Fall River Machine Co.

PRESIDENT : John S. Brayton
CLERK AND TREASURER : George H. Bush.
DIRECTORS : John S. Brayton, R. B. Borden,
Oliver S. Hawes, Philip D. Borden; H. M. Barns
of Bristol.
Annual Meeting—4th Thursday in October.

Fall River Manufactory.

PRESIDENT: John S. Brayton.
CLERK AND TREASURER: Holder B. Durfee.
DIRECTORS: John S. Brayton, Christopher Bor-
den, H. B. Durfee, Jas. M. Morton, Edw. L.
Anthony.
Annual Meeting—2d Monday in October.

Fall River Manufacturers' Mutual Ins. Co.

PRESIDENT AND TREASURER: Thos. J. Borden.
SECRETARY: Chas. S. Waring.
DIRECTORS: Thomas J. Borden, P. D. Borden,
R. B. Borden, Andrew G. Pierce, B. D. Davol,
Walter C. Durfee, D. A. Brayton, Jr., Chas. M.
Shove, S. B. Chase, Jas. C. Eddy, Joseph Healy,
Thos. E. Brayton, Edw. L. Anthony.
. Annual Meeting—1st Monday in February.

Fall River Merino Co.

PRESIDENT: Frank S. Stevens.
CLERK: Seth H. Wetherbee.
TREASURER: Joseph Healy.
DIRECTORS: Frank S. Stevens, Joseph Healy,
F. O. Dodge, Reuben Hargraves, S.H. Wetherbee.
Annual Meeting—4th Thursday in January.

Fall River Mill Supply Co.

PRESIDENT: John D. Flint.
CLERK: Jas. F. Jackson.
TREASURER: Albert O. Phillips.
DIRECTORS: John D. Flint, Joseph Shaw, Albert O. Phillips, Jas. F. Jackson, Joseph Sladding.
Annual Meeting—3d Tuesday in October.

Fall River and Providence Steamboat Co.

PRESIDENT : John S. Brayton.
CLERK AND TREASURER : David C. Lawton.
DIRECTORS : John S. Brayton, R. B. Borden, Frank S. Stevens, P. D. Borden, Robert C. Brown; H. M. Barns, Bristol.
Annual Meeting—4th Thursday in October.

Fall River Railroad.

PRESIDENT: Charles F. Choate.
CLERK : Wm. Rotch.
TREASURER: John M. Washburn.
DIRECTORS: Chas. F. Choate, Wm. Rotch, H. A. Blood, Jos. A. Beauvais, Morgan Rotch, F. L. Ames, R. W. Turner, John S. Brayton.
Annual Meeting—2d Thursday in December.

Fall River Spool and Bobbin Co.

PRESIDENT: Joseph Healy.
CLERK AND TREASURER: J. Henry Wells.
DIRECTORS: Joseph Healy, F. L. Almy, B. D. Davol, J. Henry Wells, A. J. Chace, Jas. M. Osborn.
Annual Meeting—Last Tuesday in October.

F. R., Warren and Providence R. R. Co.

PRESIDENT: Charles F. Choate.
CLERK: John S. Brayton.
TREASURER: John M. Washburn.
DIRECTORS: Chas. F. Choate, Southboro; J. S. Brayton, T. J. Borden, Fall River; Thomas Dunn, Newport; R. W. Turner, Randolph; F. L. Ames, Easton.
Annual Meeting—2d Monday in March.

Flint Mills.

PRESIDENT: John D. Flint.
CLERK: Franklin L. Almy.
TREASURER: William S. Potter.
DIRECTORS: J. D. Flint, B. D. Davol, Franklin L. Almy, Geo. W. Nowell, Reuben Hargraves, George H. Eddy, Wm. S. Potter.
Annual Meeting—1st Monday in November.

Globe Street Railway Co.

PRESIDENT : Frank S. Stevens.
CLERK : M. G. B. Swift.
TREASURER : Robert S. Goff.
SUPERINTENDENT : J. H. Bowker,
DIRECTORS : F. S. Stevens, John S. Brayton, A. J. Borden, M. G. B. Swift, Geo. H. Hawes, J. A. Beauvais, S. B. Chase.
Annual Meeting—3d Tuesday in October.

Globe Yarn Mills.

PRESIDENT : Wm. Lindsey.
CLERK AND TREASURER : Arnold B. Sanford.
DIRECTORS : Wm. Lindsey, A. B. Sanford, Andrew J. Borden, H. K. Braley; Jos. A. Beauvais, New Bedford; E. S. Draper, Hopedale; Horace M. Barns, Bristol; Oliver Ames, North Easton.
Annual Meeting—4th Thursday in April.

Granite City Soap Co.

PRESIDENT: George W. Slade.
CLERK: Andrew H. Jones.
TREASURER: Stephen B. Ashley.
DIRECTORS: Geo. W. Slade, A. H. Jones, Stephen B. Ashley, James H. Cameron; Chas. Bunker, Boston.
Annual Meeting—In October.

Granite Mills.

PRESIDENT: William Mason.
CLERK AND TREASURER: Charles M. Shove.
DIRECTORS: Wm. Mason, John S. Brayton, John P. Slade, Frank S. Stevens, Edward E. Hathaway, Robert Henry. Charles M. Shove.
Annual Meeting—4th Monday in October.

Hargraves Mills.

PRESIDENT: Reuben Hargraves.
CLERK: Milton Reed.
TREASURER: Seth A. Borden.
DIRECTORS: Reuben Hargraves, Thos. Hargraves, Milton Reed, Adoniram J. Chace, Geo. C. Silsbury, John Barlow, Seth A. Borden, John D. Flint, J. Edward Osborn, L. Lincoln; Stephen A. Jenks, Pawtucket; Geo. A. Draper, Hopedale; Wm. H. Parker, Lowell.
Annual Meeting—Last Thursday in October.

Jesse Eddy Manufacturing Co.

PRESIDENT: James C. Eddy.
CLERK: Geo. A. Mathewson.
TREASURER: Timothy E. Hopkins.
DIRECTORS: Jas. C. Eddy, Geo. A. Mathewson; Timothy E. Hopkins, Danielsonville, Conn.
Annual Meeting—2d Tuesday in April.

Kerr Thread Co.

PRESIDENT: Robert C. Kerr.
CLERK: Ina C. Davis.
TREASURER: John P. Kerr.
DIRECTORS: John P. Kerr, Robert C. Kerr, Ina C. Davis, E. Newark, N. J.
Annual Meeting—2d Tuesday in August.

Kilburn, Lincoln & Co.

PRESIDENT: Andrew Luscomb.
CLERK AND TREASURER: Leontine Lincoln.
DIRECTORS: Andrew Luscomb, Chas. H. Dring, Leontine Lincoln, Chas. P. Dring.
Annual Meeting—Last Monday in January.

King Philip Mills.

PRESIDENT: Chas. J. Holmes.
CLERK: George S. Davol.
TREASURER: Simeon B. Chase.
DIRECTORS: Chas. J. Holmes, Edwin Shaw, Henry H. Earl, Leontine Lincoln, Chas. E. Fisher, Geo. A. Ballard, S. B. Chase; Francis A. Foster, Boston; Jos. A. Beauvais, New Bedford.
Annual Meeting—Last Thursday in October.

Laurel Lake Mills.

PRESIDENT : John P. Slade.
CLERK AND TREASURER : Abbott E. Slade.
DIRECTORS : John P. Slade, S. H. Miller, John B. Whitaker, Prelet D. Conant, Leonard N. Slade, Geo. W. Nowell; J. Frank Howland, Boston; Jas. E. Easterbrooke, Swansea; Geo. R. Deardon, Somerset.
Annual Meeting—3d Tuesday in October.

Manufacturers Board of Trade.

PRESIDENT: Nathaniel B. Borden.
VICE-PRESIDENT: Bradford D. Davol.
SECRETARY AND TREAS.: C. C. Rounseville.
EXECUTIVE COMMITTEE: Nath'l B. Borden, Cyrus C. Rounseville, Jos. A. Baker, Andrew Borden.
Annual Meeting—3d Friday in February.

Manufacturers Gas Light Co.

PRESIDENT : Chas. M. Shove.
CLERK AND TREASURER : Joseph A. Baker.
DIRECTORS : Chas. M. Shove, Foster H. Stafford, James C. Eddy, Jos. A. Baker, David A. Brayton, Jr., F. H. Dwelly, C. M. Hathaway.
Annual Meeting—In July.

Massasoit Manufacturing Co.

PRESIDENT: Frank L. Palmer.
CLERK AND TREASURER; Wendell E. Turner.
DIRECTORS; Frank L. Palmer, Edward A. Palmer, Wendell E. Turner, W. H. Turner, Elisha L. Palmer.
Annual Meeting—1st Tuesday in February.

Mechanics Mills.

PRESIDENT: Thomas J. Borden.
CLERK: James M. Morton.
TREASURER: Horatio N. Durfee.
DIRECTORS: Thomas J. Borden, Job B. French, Tillinghast Records, Southard H. Miller, James M. Morton, John B. Hathaway, F. S. Stevens, John S. Brayton, Richard B. Borden.
Annual Meeting—1st Thursday in February.

Merchants Manufacturing Co.

PRESIDENT: James Henry.
CLERK AND TREASURER: Andrew Borden.
DIRECTORS: James Henry, James M. Osborn, Richard B. Borden, Robert T. Davis, Sam'l Wadington, Andrew J. Borden, A. J. Chace, Andrew Borden, Edward B. Jennings.
Annual Meeting—4th Wednesday in October.

Metacomet Manufacturing Co.

PRESIDENT : John S. Brayton.
CLERK AND TREASURER : Thomas S. Borden.
DIRECTORS : John S. Brayton, R. B. Borden, Robert C. Brown, Philip D. Borden. Clark Shove; Horace M. Barns, Bristol.
Annual Meeting—4th Wednesday in October.

Narragansett Mills.

PRESIDENT: Edward S. Adams.
CLERK AND TREASURER: James Waring.
DIRECTORS: Edward S. Adams, James Waring, George W. Nowell, Geo. H. Hawes; John H. Thompson, Prov.; Abraham Steinam, N. Y.
Annual Meeting—Last week in October.

Old Colony Railroad.

PRESIDENT: Charles F. Choate.
CLERK: John S. Brayton.
TREASURER: John M. Washburn.
DIRECTORS: Charles F. Choate, Southboro; Geo. A. Gardner, James R. Kendrick, Samuel C. Cobb, Boston; Fred. L. Ames, Easton; Chas. L. Lovering, Taunton; John J. Russell, Plymouth; John S. Brayton, T. J. Borden, Fall River; R. W. Turner, Randolph; Wm. J. Rotch, New Bedford; Thomas Dunn, Newport; Nath'l Thayer, Lancaster.
Annual Meeting—Last Tuesday in September.

Old Colony Steamboat Co.

PRESIDENT: Charles F. Choate.
CLERK: John S. Brayton.
TREASURER: John M Washburn.
DIRECTORS: Chas. F. Choate, Silas Pierce, Boston; T. J. Borden, John S. Brayton, Fall River; F. L. Ames, Easton; Nath'l Thayer, Lancaster; Cornelius N. Bliss, Leander N. Lovell, New York; Wm. Rotch, New Bedford.
Annual Meeting—1st Tuesday in June.

Osborn Mills.

PRESIDENT: Weaver Osborn.
CLERK AND TREASURER: Joseph Healy.
DIRECTORS: Weaver Osborn, John C. Milne, Joseph Healy, Edward E. Hathaway, Benjamin Hall, Jas. M. Osborn, Frank S. Stevens.
Annual Meeting—Last Tuesday in April.

Pocasset Manufacturing Co.

PRESIDENT: Horatio Hathaway.
CLERK AND TREASURER: Bradford D. Davol.
AGENT: William S. Whitney.
DIRECTORS: Samuel W. Rodman, B. R. Weld, Boston; Horatio Hathaway, Jos. F. Knowles, New Bedford; Milton Reed, B. D. Davol, A. B. Sanford, Thos. E. Brayton. Edw. L. Anthony.
Annual Meeting—Last Thursday in February.

Quequechan Mills.

TRUSTEE FOR OWNERS: Weaver Osborn.
LESSEES:

Richard Borden Manufacturing Co.

PRESIDENT: Thomas J. Borden.
CLERK AND TREASURER: Richard B. Borden.
DIRECTORS: Thos. J. Borden, Rich'd B. Borden, Jerome C. Borden, Frank S. Stevens; A. S.Covel, Boston; Edward P. Borden, Phila.; Matthew C. D. Borden, New York.
Annual Meeting—2d Tuesday in November

Robeson Mills.

PRESIDENT: Lloyd S. Earle.
CLERK AND TREASURER: C. M. Hathaway.
DIRECTORS: Wm. R. Robeson of Boston; Lloyd S. Earle, E. E. Hathaway, Nath'l B. Horton, Chas. B. Luther, C. M. Hathaway, Robt. T. Davis.
Annual Meeting—1st Monday in February.

Sagamore Manufacturing Co.

PRESIDENT: Chas. J. Holmes.
CLERK: D. Hartwell Dyer.
TREASURER: Hezekiah A. Brayton.
DIRECTORS: Chas. J. Holmes, John S. Brayton, Job M. Leonard, Jos. A. Baker, D. H. Dyer, David A. Brayton. Jr.; Moses W. Richardson, Boston; Jos. A. Beauvais, Gilbert Allen, New Bedford.
Annual Meeting—4th Wednesday in October.

Seaconnet Mills.

PRESIDENT : Leontine Lincoln.
CLERK : Milton Reed.
TREASURER : Edward A. Chace.
DIRECTORS : L. Lincoln, Wm. Beattie, Milton Reed, Reuben Hargraves, Wm. R. Warner, Sam'l Wadington ; Stephen A. Jenks, Pawtucket; George A. Draper, Hopedale ; Rufus A. Peck, Providence.
Annual Meeting—1st Thursday in November.

Shove Mills.

PRESIDENT: Charles M. Shove.
CLERK AND TREASURER: C. C. Rounseville.
DIRECTORS: Charles M. Shove, John P. Slade, Geo. A. Chace, Isaac W. Howland, George W. Slade, Fenner Brownell, Cyrus C. Rounseville ; Wm. Mason, Taunton.
Annual Meeting—In February.

Slade Mills.

PRESIDENT: William L. Slade.
CLERK AND TREASURER: Henry S. Fenner.
DIRECTORS: Wm. L. Slade, Jonathan Slade, Benjamin Hall, Frank S. Stevens, John C. Milne, Daniel Wilbur, Henry S. Fenner; Geo. W. Hills, Lawrence.
Annual Meeting—Last Tuesday in January.

Stafford Mills.

PRESIDENT: Foster H. Stafford.
CLERK AND TREASURER: Effingham M. Cock.
AGENT: Foster H. Stafford.
DIRECTORS: F. H. Stafford, Robert T. Davis, Wm. L. Slade, William Mason, Frank S. Stevens, Edward E. Hathaway, Samuel W. Hathaway, Jas. M. Osborn, Effingham M. Cock.
Annual Meeting—4th Tuesday in October.

Tecumseh Mills.

PRESIDENT: Jerome C. Borden.
CLERK AND TREASURER: Frank H. Dwelly.
DIRECTORS: Jerome C. Borden, Samuel Wadington, David T. Wilcox, John Southworth, Simeon B. Chase, George E. Hoar, George W. Nowell, Leontine Lincoln, A. J. Chace.
Annual Meeting—4th Tuesday in October.

Troy Cotton and Woolen Manufactory.

PRESIDENT: John S. Brayton.
CLERK AND TREASURER: Richard B. Borden.
DIRECTORS: John S. Brayton, Thomas J. Borden, Richard B. Borden, Andrew J. Borden; Horatio Hathaway, N. B.
Annual Meeting—1st Tuesday in February.

Union Belt Company.

PRESIDENT: Richard B. Borden.
CLERK AND TREAS: Robert N. Hathaway.
AGENT: William H. Chace.
DIRECTORS: R. B. Borden, Bradford D. Davol, Wm. H. Chace; Alphonso S. Covel, Boston; H. Martin Brown, Providence.
Annual Meeting—3d Thursday in January.

Union Cotton Manufacturing Co.

PRESIDENT: James M. Morton.
CLERK AND TREASURER: Thomas E. Brayton.
DIRECTORS: Jacob Edwards, Boston; Thomas B. Wilcox, Horatio Hathaway, Thos. M. Stetson, Jos. F. Knowles, New Bedford; John B. Anthony, Providence; F. H. Stafford, Jas. M. Morton, Thos. E. Brayton, Fall River.
Annual Meeting—4th Wednesday in October.

Wampanoag Mills.

PRESIDENT: Robert T. Davis.
CLERK AND TREASURER—Walter C. Durfee.
DIRECTORS: Robert T. Davis, W. C. Durfee, John D. Flint, Foster H. Stafford, George H. Eddy, John H. Boone, Daniel Wilbur, Lloyd S. Earle, Franklin L. Almy, Simeon B. Chase, Effingham M. Cock.
Annual Meeting—4th Monday in October.

Weetamoe Mills.

PRESIDENT: Job B. French.
CLERK: Francis B. Hood.
TREASURER: William Lindsey.
DIRECTORS: Job B. French, Josiah C. Blaisdell, Wm. Lindsey, John P. Slade, William H. Ashley, Francis B. Hood, Elisha B. Gardner, John P. Nowell, Crawford E. Lindsey.
Annual Meeting—4th Wednesday in January.

ORGANIZATION OF

National and Savings Banks.

JANUARY, 1890.

Fall River National Bank.

PRESIDENT: Guilford H. Hathaway.
CASHIER: Ferdinand H. Gifford.
ASST. CASHIER: Chas. B. Cook.
DIRECTORS: G. H. Hathaway, John P. Slade,
Richard B. Borden, Henry S. Fenner. F. H. Gifford,
Herbert Field, Chas. L. Porter, Philip H. Borden,
Wendell E. Turner.
Annual Meeting—1st Monday in January.

First National Bank.

PRESIDENT: John S. Brayton.
CASHIER: Everett M. Cook.
DIRECTORS: John S. Brayton, Hezekiah A. Brayton, D. A. Brayton, Jr., James M. Morton, Thos. E.
Brayton, Edward L. Anthony, Andrew Borden.
Annual Meeting—2d Tuesday in January.

Massasoit National Bank.

PRESIDENT: Charles M. Shove.
CASHIER: Eric W. Borden.
DIRECTORS: Chas. M. Shove, Bradford D. Davol.
Southard H. Miller, Geo. A. Chace, Nath'l B.
Borden, Benj. S. C. Gifford, Henry W. Davis, Frank
H. Dwelly, Eric W. Borden.
Annual Meeting—1st Wednesday in January.

Metacomet National Bank.

PRESIDENT: Walter C. Durfee.
VICE PRESIDENT: Frank S. Stevens.
CASHIER: George H. Borden.
DIRECTORS: Walter C. Durfee, Thos. J. Borden, Jonathan Slade, Frank S. Stevens, Geo. H. Hawes, Milton Reed, Wm. R. Warner, Wm. S. Potter, Effingham M. Cock.
Annual Meeting—2d Thursday in January.

National Union Bank.

PRESIDENT: Daniel Wilbur.
CASHIER: John T. Burrell,
DIRECTORS: Daniel Wilbur, John D. Flint, Samuel Wadington, David M. Anthony, Wm. C. Cornell, Thos. D. Covel, Stephen B. Ashley, Clark Chase, Reuben Hargraves; David P. Davis, Somerset.
Annual Meeting—2d Friday in January.

Pocasset National Bank.

PRESIDENT: Weaver Osborn.
CASHIER: Edward E. Hathaway.
DIRECTORS: Weaver Osborn, Foster H. Stafford, John C. Milne, Nathan Read, Edward E. Hathaway, Joseph Healy, Lloyd S. Earle.
Annual Meeting—1st Monday in January.

Second National Bank.

PRESIDENT: Leontine Lincoln.
CASHIER: Charles J. Holmes.
ASST. CASHIER: Wm. B. Lovell.
DIRECTORS: Leontine Lincoln, Chas. E. Fisher, Albert Winslow, Chas. J. Holmes, Arnold B. Sanford, C. C. Rounseville, Junius P. Prentiss.
Annual Meeting—2d Tuesday in January.

SAVINGS BANKS.

Fall River Savings Bank.

PRESIDENT: Crawford E. Lindsey.
SECRETARY: Newton R. Earl.
TREASURER: Charles A. Bassett.
BOARD OF INVESTMENT: Guilford H. Hathaway, Robert C. Brown, James C. Eddy, Henry C. Hawkins, Robert Henry.
Annual Meeting—2d Wednesday in January.

Citizens Savings Bank.

PRESIDENT: John C. Milne.
SECRETARY: Henry H. Earl.
TREASURER: Edward E. Hathaway.
BOARD OF INVESTMENT: Weaver Osborn, John C. Milne, Lloyd S. Earle, S. W. Hathaway, Marcus G. B. Swift.
Annual Meeting—2d Monday in June.

Fall River Five Cents Savings Bank.

PRESIDENT: Walter C. Durfee.
SECRETARY: John P. Slade.
TREASURER: Charles J. Holmes.
BOARD OF INVESTMENT: W. C. Durfee, Samuel M. Brown, Edwin Shaw, L. Lincoln, Chas. E. Gifford.
Annual Meeting—Saturday before 1st Monday in December.

Union Savings Bank.

PRESIDENT: Andrew J. Borden.
SECRETARY: Abraham G. Hart.
TREASURER: Jerome C. Borden.
BOARD OF INVESTMENT: Andrew J. Borden, Wm. W. Stewart, Geo. W. Dean, A. Homer Skinner, Thomas D. Covel.
Annual Meeting—4th Wednesday in May.

CO-OPERATIVE BANKS.

Fall River Co-Operative Bank.

Chartered, 1888.   Authorized Capital, $1,000,000
PRESIDENT: John Barlow.
VICE-PRESIDENT: Eric W. Borden.
SECRETARY: George O. Lathrop.
TREASURER: Rodolphus N. Allen.
DIRECTORS: Henry W. Davis, Frank H. Dwelly,
William J.Wiley, John Duff, Henry Waring, Enoch
J. French, Arba N. Lincoln, George N. Durfee,
Charles F. Tripp, Alfred H. Hood, Henry C. Hampton, Richard Rushton, Clarence A. Brown.
Annual Meeting—2d Wednesday in October.
Monthly Meeting—2d Wednesday of each month

Peoples Co-Operative Bank.

Chartered, 1882.—Authorized Capital, $1,000,000.
PRESIDENT: Milton Reed.
VICE PRESIDENT: Samuel M. Brown.
SECRETARY AND TREASURER: Samuel Hadfield.
DIRECTORS: Charles E. Mills, Joseph Clifton,
Edward S. Adams, Edward A. Mott, Reuben Hargraves, John H. Estes, Patrick Kiernan, Abner P.
Davol, Samuel Hyde, Seth R. Thomas, Owen Durfee, J P. Prentiss, Fred. O. Dodge, Seth A. Borden,
John H. Hadfield.
Annual Meeting—2nd Monday in November.
Monthly Meeting—3d Wednesday of each month.

Troy Co-Operative Bank.

Chartered, 1880.—Authorized Capital, $1,000,000.
PRESIDENT: Jerome C. Borden.
VICE PRESIDENT: Cyrus C. Rounseville.
SECRETARY AND TREASURER: Chas. B. Cook.
DIRECTORS: John M. Young, Albert F. Dow,
James E. O'Connor, Thos. D. Covel, Nathaniel B.
Borden, James E. McCreery, Joseph Bowers, A.
Homer Skinner, Walter R. Woodman, James H.
French, John A. Sanford.
Annual Meeting—2nd Monday in November.
Monthly Meeting—3d Tuesday of each month.

B. M. C. Durfee Safe Deposit and Trust Co.

Chartered, 1887.   Authorized Capital, $500,000.
Capital Paid in, $200,000.

PRESIDENT: John S. Brayton.
VICE-PRESIDENT: Thomas E. Brayton.
SECRETARY AND TREASURER: Arthur W. Allen.
DIRECTORS: John S. Brayton, Thomas E. Brayton, Edward L. Anthony, Byron W. Anthony, George A. Ballard, Philip D. Borden, Andrew J. Borden, Andrew Borden, Hezekiah A. Brayton, David A. Brayton, Jr., James M. Morton.
Annual Meeting—2d Tuesday in January.
Discount Day—Every Day.
MANAGER OF SAFE DEPOSIT DEPT.—David S. Brigham.

Banks of the City of Fall River, Mass., from Official Reports, Dec. 11, 1889.

Name.	Estab.	President.	Cashier.	Capital.	Surplus & Int.	Disc't Day.
Fall River Nat'l Bank,	1825	G. H. Hathaway,	F. H. Gifford,	400,000	133,438	Mon.
National Union Bank,	1830	Daniel Wilbur,	John T.Burrell,	200,000	41,555	Fri.
Massasoit Nat'l Bank,	1846	Chas. M. Shove,	Eric W.Borden,	200,000	173,905	Wed.
Metacomet Nat'l Bank,	1853	W. C. Durfee,	Geo.H.Borden,	500,000	198,762	Mon. } Thu. }
Pocasset Nat'l Bank,	1854	Weaver Osborn,	E. F. Hathaway,	200,000	89,132	Tues.
Second National Bank,	1857	Leontine Lincoln	C. J. Holmes,	150,000	89,099	Thurs.
First National Bank,	1864	John S. Brayton,	E. M. Cook,	400,000	196,599	Daily.
				2,050,000	922,490	

Savings Banks of Fall River, Mass., from the Official Reports, Jan., 1890.

Name.	Incor.	Treasurer.	Deposits.	Depos'rs.	Disc't Day.	Dividends.	
Fall River Savings Bank,	1828	C. A. Bassett,	5,866,751.92	11,557	Tues.	Apr.	Oct.
Citizens' Savings Bank,	1851	E. E. Hathaway,	3,166,802.41	5,253	Fri.	June.	Dec.
F.R.FiveCent Sav'gs Bank,	1856	C. J. Holmes,	2,298,098.00	8.007	Mon.	June.	Dec.
Union Savings Bank,	1869	J. C. Borden,	806,006.21	1,778	Fri.	Nov.	May.
			12,132,658.54	26,595			

➤✳Fire Alarm.✳◄

12, Cor. Central & Davol.
121, Cor. Green and Elm.
123, Cor. N. Main & Cherry.
124, Depot at Steamboat Dock.
125, Cor. Cedar & Durfee.
126, Davol, foot of Cedar St.
127, Massasoit Mfg. Co.*
13, Cor. N. Main & Turner.
131, Cor. N. Main & Brownell.
132, Mechanics Mills
134, Weetamoe Mills
135, Cor. George & Lindsey.
136. Cor. Davol & Pierce.

14, Narragansett Mills.*
141, Cor. N. Main & Langley.
142, B. C. Mfg. Co.,Mill No. 1.
143, B. C. Mfg.Co.,Mill, No. 2.
145, Sagamore Mill, No. 1.*
146, N. Main, near B. C. Mills.
147, Sagamore Mill, No. 2.*
148, Steep Brook Corners.

2, Cor. Broadway & S. Main.
21, Slade Mills.*
23, King Philip Mills.*
24, Osborn Mill, No. 2.*
25, Osborn Mill, No. 1.*
26, Cor. S. Main and Osborn.
27, Cor. Bay and Chace.
28, S. Main, opp. Shove Mills.
29, Shove Mills.*
212, Globe Yarn Mills.*
213, Laurel Lake Mills.*
214, Globe Yarn Mill, No. 3.*
216, Cor. Bay and Sprague.
231, Cor. S. Main & King Philip
272, Conanicut Mills.*

3, Cor. Union and Spring.
31, Cor. Canal and Ferry.
32, Cor. Canal and Anawan.
34, American Print Works.*
35, Fall River Iron Works.*
36, Ferry Street Depot.
37, American Linen Mills.*
38, Cor. William and Almond.
39, Cor. Mulberry & Division.
311, Metacomet Mills.*
312, Fall River Manufactory.*
313, Quequechan Mills.*
314, Pocasset Mills.*

4,Cor. Rodman and Third.
41, Cascade Hose House.
42, Cor. Second and Branch.
43, Cor. Park and Ridge.
45, Gas Works, Hartwell st.

46, Davol Mills.*
47, Tecumseh Mill, No. 1.*
48, Robeson Mill.
49, Cor. John and Morgan Sts

5, Cor. E. R. Way & Rodman
51, Tecumseh Mill, No. 2.*
52, Richard Borden Mills.*
53, Chace Mills.*
54, Cor. E. R. Way & Stafford.
56, Cor. Six and E. R. Ways.
57, Fall River Bleachery.*
58, Cor. Rodman & Warren.
59, Cor. Tecumseh &E.R.Way
571, Cor. Staf. Road & Lawton.
581, Barnaby Mills.*
582, Spool & Bobbin Mill*.
583, Davis's Pork Factory.*

6, Wamsutta Woolen Mills.*
61, Union Mills.*
62, Durfee Mills.*
63, Crescent Mills.*

7, City Hall.
71, Troy Mills.*
72, Cor. Pleasant and Sixth.
73, Cor. Pleasant & Twelfth
74,Granite Mills.*
75, Merchants Mills.*
79, Cor. Pleasant & Quarry.

8, Stafford Mills.*
81, Wampanoag Mills.*
82, Cor.Alden, near Flint Mill.
83, Flint Mills.*
84, Barnard Mills.*
85, Cor. Webster & Pleasant.
86, Cor, Haffard & County.
87, Cor. Bedford and Covel.
89, Cor. Pleasant & Mason.
812, Seaconnet Mills.*
813, Cor. Pleasant & Barlow.
814, Hargraves Mills.*
815, Cornell Mill.*

9, Cor. Rock & Franklin.
91, Cor. Bedford & Oak.
92, Cor. Pine & Grove.
93, Cor. Orange & Bedford.
94, Cor. Maple & Rock.
95, Cor. Highl'd Av.& French.
96, Cor. High & Cherry.
97, Cor. Maple & Hanover.
98, Cor. N. Main & Lincoln av
912, Cor. Walnut and Grove.

www.ingramcontent.com/pod-product-compliance
Lightning Source LLC
Chambersburg PA
CBHW030614290326
41930CB00050B/1343